Rocks and Minerals

Geology of Pacific Northwest

Written and compiled by
Autumn Christensen, Reanna Camp-Witmer,
and Shannon Othus-Gault

with additional contributions from Sheila Alfsen, Andrew Frank,
Bill Orr, and Mariah Tilman

Rocks and Minerals: Lab Manual
ISBN: 978-1-943536-61-0
Edition 2 Fall 2019
© 2019 Chemeketa Community College. All rights reserved.

Chemeketa Press

Chemeketa Press is a nonprofit textbook publisher that works directly with faculty authors to make affordable, effective, engaging, and accessible textbooks. We bring the passion and enthusiasm of your favorite professor to the page, through agile publishing methods that will change the industry. We do this for our students and yours, nationwide—because a textbook should open the door, not close it.

To learn more, visit www.chemeketapress.org.

Publisher: David Hallett
Director: Steve Richardson
Managing Editor: Brian Mosher
Instructional Editor: Stephanie Lenox
Design Editor: Ronald Cox IV
Cover Design: Brandi Harbison
Interior Design and Layout: Brice Spreadbury, Cierra Maher
Cover Photo: Shallow Focus Photography of Gravels, by Nick Nice is in the public domain (https://unsplash.com/photos/zwjSCTltiZU).

Acknowledgments

Lab 2: Rock Cycle in Chocolate, is licensed under a Creative Commons CC-BY-ND-SA 3.0 license (http://serc.carleton.edu/NAGTWorkshops/intro/activities/23590.html).

Lab 3, Figure 1. "Chain Comparison" by Cierra Maher is a derivative work of images in the Public Domain (https://commons.wikimedia.org/wiki/File:EmperorSeamounts.jpg & https://commons.wikimedia.org/wiki/File:Hawaiian_seamount_chain.jpg).

Lab 3, Figure 2. "Plate Boundry Map." This map is under copyright and is used with permission of Dale S. Sawyer and Rice University (http://plateboundary.rice.edu/downloads.html).

Lab 3, Figure 3. "Plate Boundry Map." This map is under copyright and is used with permission of Dale S. Sawyer and Rice University (http://plateboundary.rice.edu/downloads.html).

Lab 3, Figure 4 . "Earthquake Depth and Distribution Map." This map is under copyright and is used with permission of Dale S. Sawyer and Rice University (http://plateboundary.rice.edu/downloads.html).

Lab 3, Figure 5. "Age of the Ocean Floor." This map is under copyright and is used with permission of Dale S. Sawyer and Rice University (http://plateboundary.rice.edu/downloads.html)

Lab 7: M&M Magma Chamber, is licensed under a Creative Commons CC-BY-ND-SA 3.0 license (http://serc.carleton.edu/NAGTWorkshops/petrology/teaching_examples/24646.html).

Printed in the United States of America.

Contents

Labs
Lab 1: Rock Cycle - What Geologists See — 1
Lab 2: Rock Cycle in Chocolate — 5
Lab 3: Plate Tectonics — 17
Lab 4: Mineral Properties — 29
Lab 5: Rock Forming Minerals — 35
Lab 6: Economic Minerals — 37
Lab 7: M&M Magma Chamber — 45
Lab 8: Igneous Rocks — 53
Lab 9: Igneous Rocks of Oregon — 57
Lab 10: Sedimentary Rocks — 61
Lab 11: Sedimentary Rocks of Oregon — 65
Lab 12: Metamorphic Rocks — 69
Lab 13: Metamorphic Rocks of Oregon — 73
Lab 14: Fossils — 75
Lab 15: Discovering Plate Boundaries — 79

Field Trips
Field Trip 1: Downtown Salem Walking Tour — 81
Field Trip 2: Oregon Coast — 89
Field Trip 3: Shellburg and Henline Falls — 95

Mineral Tables — 39

Lab 1
Rock Cycle - What Geologists See

Name _____   Date _____

Purpose: An introduction to the rock cycle and the scientific method.
Materials: Rock sample(s) provided by your instructor.
Instructions:

Part 1: Answer the following questions about your rock sample. Do not be afraid to use sketches.

1. List at least five things to describe your rock. If you have taken a geology class before, do not use geologic terms. Describe it as you would to someone completely oblivious to geology.

2. Suggest an origin for your rock. How do you think it formed? Use your observations from the previous question to support your idea.

3. Suggest an experiment you could run to test your hypothesis.

Part 2: Find other people who you think have rocks that share something in common with yours. This may be that they have a similar appearance or feature, or that you think they share a common formation history. Ideally there should be 3 large groups. Answer the following questions:

1. List at least five things to describe your rocks as a group (features they all share in common). If you have taken a geology class before, do not use geologic terms. Describe them as you would to someone completely oblivious to geology.

2. Suggest a shared origin for your rocks. How do you think they formed? Use your observations from the previous question to support your idea.

3. While your rocks may seem similar (which is the why you're now grouped together) odds are they have some differences. What are some key differences between your samples?

Part 3: Split and rearrange your groups to become even more specific as to what rocks share something in common (or what rocks do not share something in common). Ideally there should now be 6 groups. Answer the following questions:

1. List at least five things to describe your rocks as a group (they all share in common). If you have taken a geology class before do not use geologic terms. Describe them as you would to someone completely oblivious to geology.

2. Suggest a shared origin for your rocks. How do you think they formed? Use your observations from the previous question to support your idea. Try to be more specific with your origin than in part 2.

3. While your rocks may seem similar (which is the why you're now grouped together) odds are they have some differences. What are some key differences between your samples? How would you explain these differences in the context of your suggested origin?

Lab 2
Rock Cycle in Chocolate

Name _____   Date _____

Purpose: An introduction to the rock cycle
Materials:

- Chocolate (dark and white)
- Aluminum foil
- Wax paper
- Hot plate
- Piece of rock countertop
- Dixie cups with cold water (ideally sitting in an ice bath)
- Hand lens
- Small glass jar with lid
- Two Plexiglas sheets

- Clamp (for metamorphism)
- Gloves
- Toothpick
- Aphyric basalt and/or vesicular basalt (Sample number _____)
- Granite (Sample _____)
- Quartz sandstone (Sample _____)
- Quartzite (Sample _____)
- Gneiss (Sample _____)
- Conglomerate (Sample _____)

Instructions:

Stage 1: Making Chocolate Magma

Make a hypothesis. We are going to be putting white and dark chocolate on the hot plate.

- Do you think the two types of chocolate will melt in the same manner? Describe what you think will happen:

Test your hypothesis. Cover the surface of the hot plate with aluminum foil and turn the hotplate to ~2 (no hotter!). On another piece of aluminum foil, place a small chunk (1 cm³) of dark chocolate and a similar sized piece of white chocolate next to each other, then place on the hot plate. You will need to use a toothpick to stir the chocolate on the foil to help the entire chunk melt. **BE CAREFUL NOT TO TEAR THE FOIL WITH THE STICK** — we don't want to spill chocolate on the hot plate. Try to keep the white and dark chocolate separate. Remember to treat both samples the same (same amount of heat, smearing, time on the hot plate, etc). Make notes while chocolate is melting, but keep an eye on it and move on to Stage 2 as soon as the chocolate is fully melted. If you leave it on too long it will burn and the experiment will not work.

Before moving on to the next step, mix the white and dark chocolate together. This is the result of two magmas mixing before they solidified.

- Describe the results of your experiment. What happened to the chocolate? Describe how both types of chocolate melted. Did both types of chocolate melt the same? Be sure to include viscosity ("thickness" or "stickiness") and how long it took to melt each sample type (relative to each other).

- Expand your conclusions. What does this experiment tell you about how substances melt? Can you extrapolate the information you just gathered to other systems (especially rocks)?

Stage 2: Making Igneous Chocolate

Turn off the hot plate.

- Now that we have made chocolate "magma," it must cool to become an igneous "rock." We will cool the magma in two ways, fast and slow. What do you think will happen? What differences do you think will result?

- Bring out the polished rock slab (countertop) or tile and the beaker of ice water. Pour a small amount (~50% of your total sample) on the polished rock slab or tile. Take the rest of your sample and pour it into the ice water. Which will cool faster? Why?

While your samples are cooling, answer the following questions (You may also want to move ahead to the next section while you wait for the slab sample to cool, just make sure you remove the chocolate from the ice water so it does not dissolve):

- What geologic environment would allow magma to cool slowly? What might cause magma to cool more rapidly? Don't forget to think relatively here. Rock freezes between 750 and 1000°F, chocolate freezes (solidifies) at a much lower temperature than rock!

- Look at your granite and basalt samples. Describe the differences between these two igneous rocks. You can use the hand lens to look up close at each sample.

- Which of these two rocks (granite or basalt) do you think cooled faster? What clues in the rock give it away?

Now look at your experimental samples. **AVOID TOUCHING THE SAMPLES WITH YOUR FINGERS** as our newly-formed chocolate will melt much faster than it did originally (this is why it is hard to make good chocolate). Using the knife or scoop, retrieve the sample from the ice water and place it on the slab next to the other sample. Using the knife, cut each of the samples in half.

- Using a hand lens, look closely at the textures of the two samples. Are there any differences?

- Now the fun part. The texture differences can be pretty minor in these samples. It is easier to tell the differences using your teeth. Carefully grind a small piece of each sample in your teeth or between your fingers. Can you detect any differences now? Describe them.

- Now look the granite and basalt samples. What do you think caused the differences in texture (grain size) between these two rocks?

- The texture differences between the two rock samples are significantly greater than the chocolate samples. How do you explain this?

Stage 3: Making Chocolate Sediment

At this stage, we're going to have to cheat a little because these processes are hard to duplicate in the lab (in fact, reproducing natural conditions in the lab is often one of the hardest parts of doing experiments!)

The formation of sediment requires processes that turn a few big objects into a lot of little objects. Use the knife to cut some of the chocolate into flakes and chips.

- You now have a pile of a various types and compositions of material. In geology this is called sediment. What will it take to turn this pile of chocolate sediment into a cohesive mass (a rock)?

- Place all the sediment on a piece of aluminum foil and wrap the sediment up in the foil. Place the foil envelope between the two pieces of Plexiglas and put the heaviest rock available on top of the Plexiglass. Why put the rock on top of the sample? What geologic process does this represent?

- While the chocolate sample is sitting under a rock, use the hand lens (magnifying glass) to look up close at your sandstone sample. This is a sedimentary rock that formed from sediment produced from rocks like granite. Describe the differences between granite and sandstone.

- The light-gray, almost transparent grains in granite are quartz. Can you identify quartz in the sandstone? How is the quartz different between the two rocks?

- Estimate the percentage of the granite sample that is quartz. Do the same for the sandstone. Are there any other minerals present in either sample? Describe (or name) them. Assuming that the sandstone is formed from sediment from the granite, is anything missing in the sandstone? What happened to that material?

- We've cheated here by manually creating sediment by breaking down the large piece of chocolate into shavings and chunks. How do you think this happens in nature with rocks? What processes do you think might be involved?

- Take the knife and cut a few more shards of chocolate from the original piece. On the left side below (under "before"), draw one or two of the shards. Now put the shards in the small bottle and close the lid. Shake the jar vigorously for 1 minute. Look at the shards again. What has happened to them? Draw what the shards look like now on the right (under the word "after").

  BEFORE   **AFTER**

- What would happen if you shook the jar for 5 minutes? An hour?

- Compare the crystals in granite and sandstone. Do you see the results of this process in the sedimentary rock? What geologic process causes this effect? Comparing the shape of the crystals, and considering that quartz is much harder than chocolate, estimate how long this process might have gone on.

- Now remove the rock from the sample and unwrap the foil envelope. What has happened to the sediment? Is it hard? Is it soft? What are *two* things that would make this "rock" more cohesive?

In geology, this process is called **lithification**. This is how sediment is converted to sedimentary rock, and is an essential process in the formation of much of the Earth's crust and most of the rocks you see around you.

NOTE: Set aside a small portion of your sedimentary chocolate "rock" for later comparison.

Stage 4: Metamorphic Chocolate

Take the sedimentary rock you've just created and put it on wax paper. The rock may come off in pieces, so just stack the pieces on top of one another. Scrape or break some more chocolate on the sedimentary rock to increase the volume of your sample, and be sure to use different colors. Wrap the chocolate in the wax paper to create as small and tight an envelope as you can.

- In order to make a sedimentary rock we used the weight of a rock on top of the sample. Predict what will happen if we apply more pressure. Consider what will happen on the scale of each grain and on the scale of the sample as a whole.

Place the envelope between the Plexiglass sheets and, using the clamp; squeeze the sample as tightly as you can. While it is pressing, answer the following questions:

- Why are we using a clamp now, rather than just adding another rock to the sedimentary apparatus? What does this say about the differences between the sedimentary and metamorphic environments?

- Look at the sandstone and quartzite samples. These rocks are composed of the same material and have the same composition. Using your hand lens, examine the details of both rocks. Describe the difference between them. Be sure to include the differences in the shape of the grains and how the grains contact one another.

- Quartzite is metamorphosed sandstone. What does that mean? Does this fit in with your hypothesis you described in question 4.1? Specifically, what matches your hypothesis? What doesn't?

- Remove the clamp from your experimental sample and unwrap the wax paper envelope. What has happened to the sample? Compare this sample to the sedimentary sample you set aside, using both words and a drawing.

- Look at the gneiss sample, another metamorphic rock. Consider how much additional pressure was required to form the metamorphic sample you just created over that required to create a sedimentary sample. With respect to depth, where do you think a rock like gneiss might form?

- We have used pressure to metamorphose this sample. Is there another tool we could use to change the shape and texture of the sample? What is it? How could we include that in our experimental design?

- What do you think would happen if we added even more pressure to this sample? What if we used the tool you discussed in question 4.6? What would eventually happen?

Stage 5: Short-circuiting the Rock Cycle

- Imagine taking the metamorphic sample you just created and cutting it into small pieces (like you did with the igneous sample) and compress it with the weight of a rock. What type of rock would this represent?

- Thinking geologically, how could this happen? What would need to occur in order for the metamorphic sample to be turned into sediment? Considering your response to question 4.7, and given that the rate of many geologic movements are in cm per year, how long do you think it might take for this to occur (give a rough estimate)?

- What would happen if you took the igneous sample you created and squeezed it with the clamp? What type of "rock" would result? Would it look the same as the sample you created in the experiment?

Stage 6: The Story of the Rocks

Examine the conglomerate sample. In a concise paragraph (with proper sentences and grammar) write the "life story" of this rock in the context of the rock cycle. Start with the "fragments that make up the rock, how might they have formed, then discuss how they came to be incorporated into the current rock, list any phases that you think this rock may have passed through to end up the way it is now.

Lab 3
Plate Tectonics

Name _____   Date _____

Purpose: To identify and distinguish between the various plate boundaries and calculate plate velocities.
Materials:

1. Maps
   a. General plate boundary Map
   b. Seafloor Ages Map
   c. Seismology Map
   d. Hawaiian Islands/Emperor Seamount Map
2. Calculator
3. Colored Sticky dots
4. Colored pencils

Instructions: Observe the maps (figures 1–5) and answer the questions below.

PART 1: IDENTIFY PLATE BOUNDARY FEATURES AND TYPES

1. Using the General Plate Boundary Map (figure 2, page 25) draw arrows representing the direction of relative plate movement on both sides of the boundaries at the following locations:

   - North American/Pacific (south of Alaska AND along U.S. west coast near California)
   - Pacific/Nazca
   - Nazca/South American
   - Caribbean/North American
   - Cocos/North American
   - South American/African
   - North American/African
   - North American/Eurasian
   - African/Eurasian (near Mediterranean)
   - Australian-Indian/Pacific
   - Australian-Indian/Eurasian (northern India)
   - Australian-Indian/Eurasian (Indo-Pacific region)
   - Philippine/Pacific
   - Philippine/Eurasian

2. Label the following trenches:
   - Aleutian Trench
   - Puerto Rico Trench
   - South Sandwich
   - Tonga and Kermadec Trenches
   - Mariana Trench
   - Ryukyu Trench

3. Label the following Mid Ocean Ridges:
   - Mid Atlantic Ridge (MAR)
   - East Pacific Rise (EPR)
   - Mid Indian Ridge (MIR)

4. Note that most mid ocean ridges have a zig-zag appearance. Explain this pattern.

5. What is the name of the plate offshore (west) of the North American Pacific Northwest?

    a. _____

    b. Sketch and label this on your map.

6. What is the name of the plate between the South American and Antarctic plate?

    a. _____

    b. Sketch and label this on your map.

7. Sketch and label the location of the East African Rift.

8. Sketch the extent of Cascade Range mountains using something like the caret symbol "^".

9. Sketch the approximate extent of the Andes Mountains using the same symbol.

10. Sketch the approximate extent of the Himalayan Mountains using the same symbol.

11. Lightly shade convergent boundaries in green.

12. Lightly shade divergent boundaries in red.

13. Lightly shade transform boundaries in blue.

PART 2: VOLCANISM AND MAGMA PRODUCTION AT PLATE BOUNDARIES

1. Describe the compositional differences between andesite and basalt.

2. The 21 volcanoes listed in the chart below represent a reasonable global distribution of volcanoes and their primary magma type (basaltic and andesitic). Using colored sticky dots, plot the following volcanoes on figure 3 (page 26) as accurately as possible according to the latitude and longitude coordinates given. Use one color dot for andesitic volcanoes, and a different color dot for basaltic volcanoes. Indicate the number designation of the volcano clearly on each sticky dot.

Table 1: Types of Igneous Rocks

#	Name	Type	Latitude	Longitude
1	Krakatoa	Andesite	6° 06' S	105° 25' E
2	Chimborazo	Andesite	1° 28' S	78° 49' W
3	Juan Fernandez	Basalt	33° 39' S	78° 49' W
4	Irazu	Andesite	9° 59' N	83° 51' W
5	Galapagos	Basalt	0° 45' S	90° 58' W
6	Katmai	Andesite	58° 16' N	155° 00' W
7	Espiritu Santo	Andesite	15° 18' S	166° 53' E

Table 1: Types of Igneous Rock (cont'd)

#	Name	Type	Latitude	Longitude
8	Haleakala	Basalt	20° 42' N	156° 15' W
9	Soufrière Hills	Andesite	16° 43' N	62° 11' W
10	Ruapehu	Andesite	39° 17' S	175° 34' E
11	Bardarbunga	Basalt	64° 38' N	17° 32' W
12	Santa Maria	Andesite	14° 45' N	91° 33' W
13	Pinatubo	Andesite	15° 08' N	120° 21' E
14	Puyehue	Andesite	45° 35' S	72° 07' W
15	Easter Island	Basalt	27° 7' S	109° 22' W
16	Aconcagua	Andesite	32° 39' S	70° 00' W
17	Eniwetok	Basalt	11° 30' N	162° 20' E
18	St. Helens	Andesite	46° 11' N	122° 12' W
19	Fuji	Andesite	35° 21' N	138° 44' E
20	Kilimanjaro	Andesite	03° 05' S	37° 21' E
21	Nevado del Ruiz	Andesite	04° 54' N	75° 19' W

Carefully examine the locations of these volcanoes by magma type.

3. Andesitic volcanoes are located primarily at what type of plate boundary?

4. Note the locations of basaltic volcanoes. Some are found on plate boundaries while others are not.

   a. Those that are located at plate boundaries are found at what type?

   b. What do we call the type of feature where the "intra" (within) plate basaltic volcanoes are located? Describe this feature.

5. What type of plate boundary does not appear to produce any volcanism?

6. At the boundary of the Indian/Eurasian plate, what feature exists **AND** why do you suppose there is virtually no volcanism here?

PART 3: EARTHQUAKES AT PLATE BOUNDARIES

Observe the earthquake depth and distribution map (figure 4, page 27) and compare it to your plate boundary and volcanic activity maps. Answer the following questions

1. Most earthquakes worldwide are of what depth? (Circle one)

    a. Shallow   b. Intermediate   c. Deep   d. Very deep

2. In general, what depth of earthquakes do you see associated with each of the following?

    a. Mid-Ocean Ridges _____

    b. Intra-plate regions _____

    c. Basaltic volcanism _____

    d. Andesitic volcanism _____

3. The deepest earthquakes are found primarily at what type of plate boundaries? (Circle one)

    a. Divergent   c. Convergent – collisional

    b. Convergent – subduction   d. Transform

4. Explain your answer above. Why do you think the deepest earthquakes are found here?

5. Look closely the Pacific Northwest and the depth of earthquakes here. Is this characteristic of a subduction zone? Explain this seeming contradiction.

PART 4: RATES OF PLATE MOVEMENT

Observe the map of the Hawaiian Island and Emperor Seamount Chains (figure 1, page 22) and answer the questions below.

1. What type of magma is produced at Hawaii?

2. Hawaii is both volcanically and seismically active. Is it due to its proximity to a plate boundary? If not, explain the presence of its extensive volcanism.

3. Look up the minimum and maximum ages of all of the Hawaiian islands shown. Write the age range in the corresponding spaces on the map of Hawaii (figure 1).

4. Why do the older islands decrease in size the further they are from the hotspot?

5. Looking specifically at the Hawaiian Islands, indicate the direction of plate movement.

6. Now view the orientation of the Emperor Seamount Chain. Assuming they were created from the same source, what does this tell us about the movement of the Pacific Plate over the past few 10s of millions of years?

7. Why do you think the Hawaiian Islands are above water and the Emperor seamounts are below water?

8. What are the minimum and maximum ages of the island of Kauai?
    a. Minimum: _____ Ma = _____ years
    b. Maximum: _____ Ma = _____ years

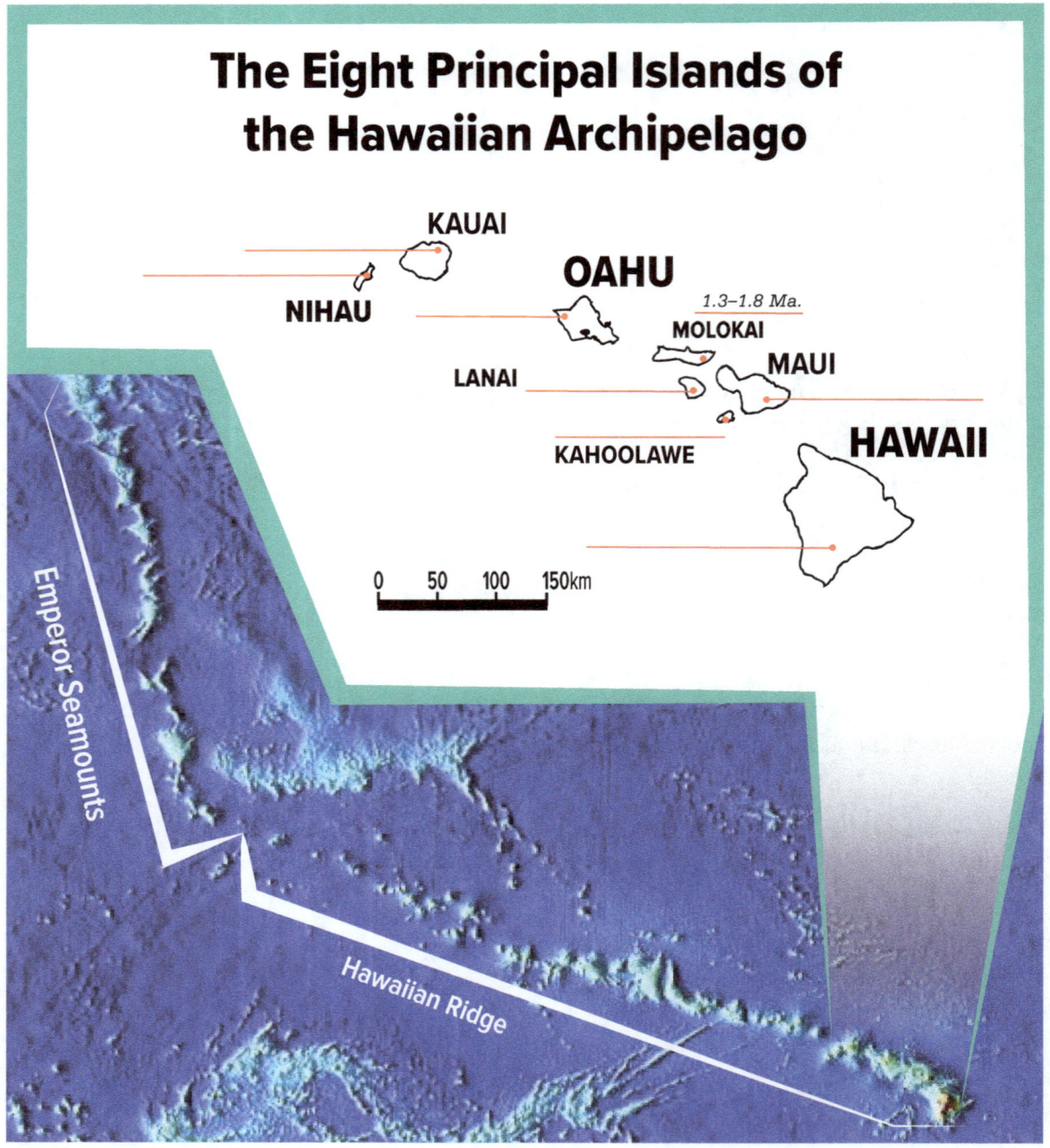

Figure 1: Hawaiian Island and Emperor Seamount Chain comparison.

9. Determine the distance from the hot spot to the center of Kauai.
   a. Distance: _____ kilometers
   b. (Convert): _____ centimeters

10. Using the data from questions above, calculate the maximum and minimum rates (velocities) of Pacific plate movement in centimeters per year.
    a. Maximum velocity: _____ cm/yr
    b. Minimum velocity: _____ cm/yr

Observe the map of the age of the ocean floor (figure 5) and answer the following questions:

11. What type of rock dominates the ocean floor? Summarize the origin, texture, and composition.

12. What are the minimum and maximum ages of the ocean floor?
    a. Minimum: _____
    b. Maximum: _____

13. Explain how do these ages compare with age ranges of continental crustal rocks?

14. Explain where you find the youngest rock on the ocean floor.

15. Explain where you find the oldest rock on the ocean floor.

16. Compare the ages of rock surrounding both the Mid-Atlantic Ridge and East Pacific Rise. Compare the spreading rates of each location (i.e., which one is spreading much faster?)

Lab 3: Plate Tectonics

PLATE BOUNDARY MAP
This map is from Dietmar Mueller, Univ. of Sydney

This map is part of "Discovering Plate Boundaries," a classroom exercise developed by Dale S. Sawyer at Rice University (dale@rice.edu). Additional information about this exercise can be found at http://terra.rice.edu/plateboundary .

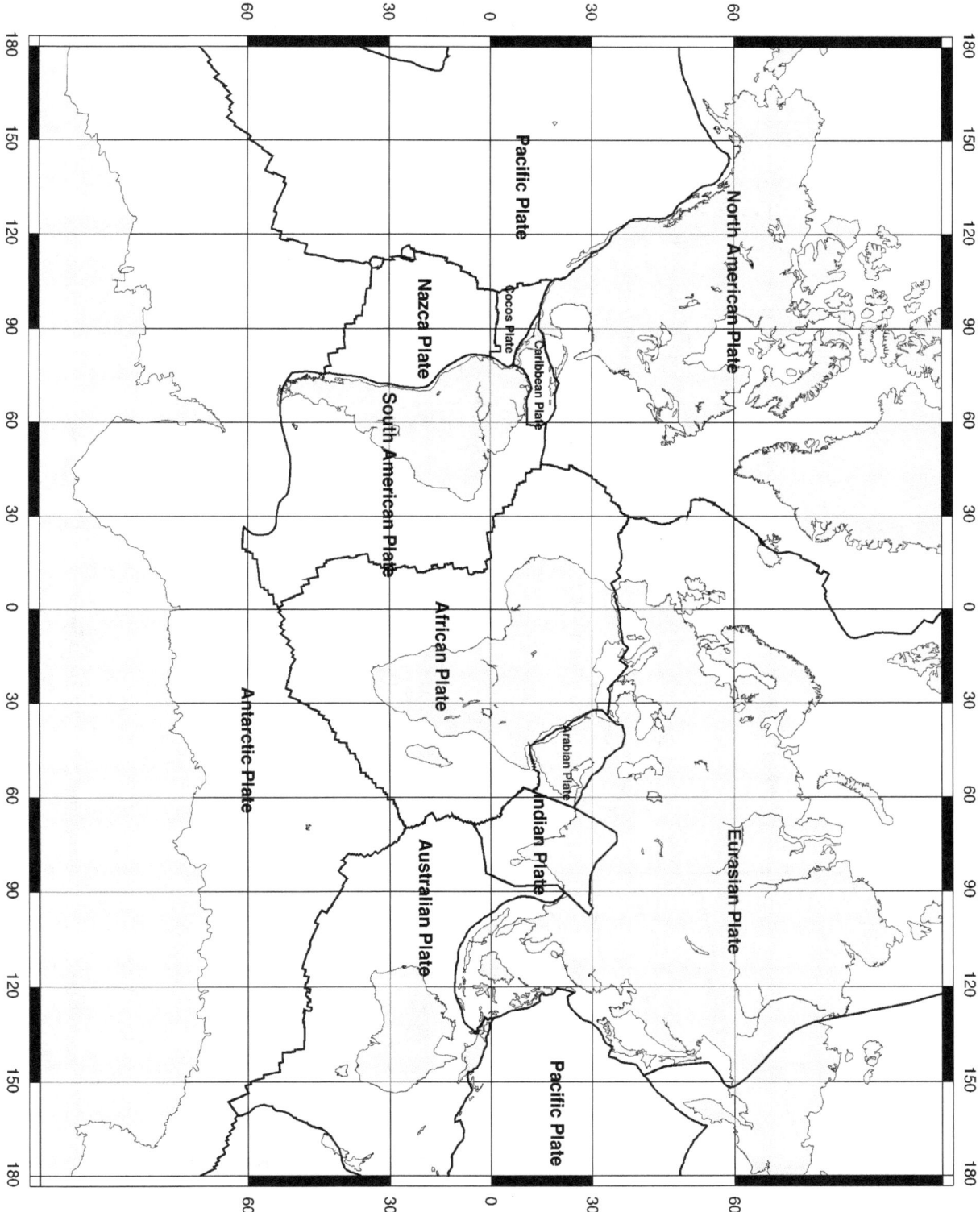

Figure 2: Plate boundary map. This map is used courtesy of Dale S. Sawyer and Rice University (http://plateboundary.rice.edu/downloads.html).

PLATE BOUNDARY MAP

This map is from Dietmar Mueller, Univ. of Sydney

This map is part of "Discovering Plate Boundaries," a classroom exercise developed by Dale S. Sawyer at Rice University (dale@rice.edu). Additional information about this exercise can be found at http://terra.rice.edu/plateboundary .

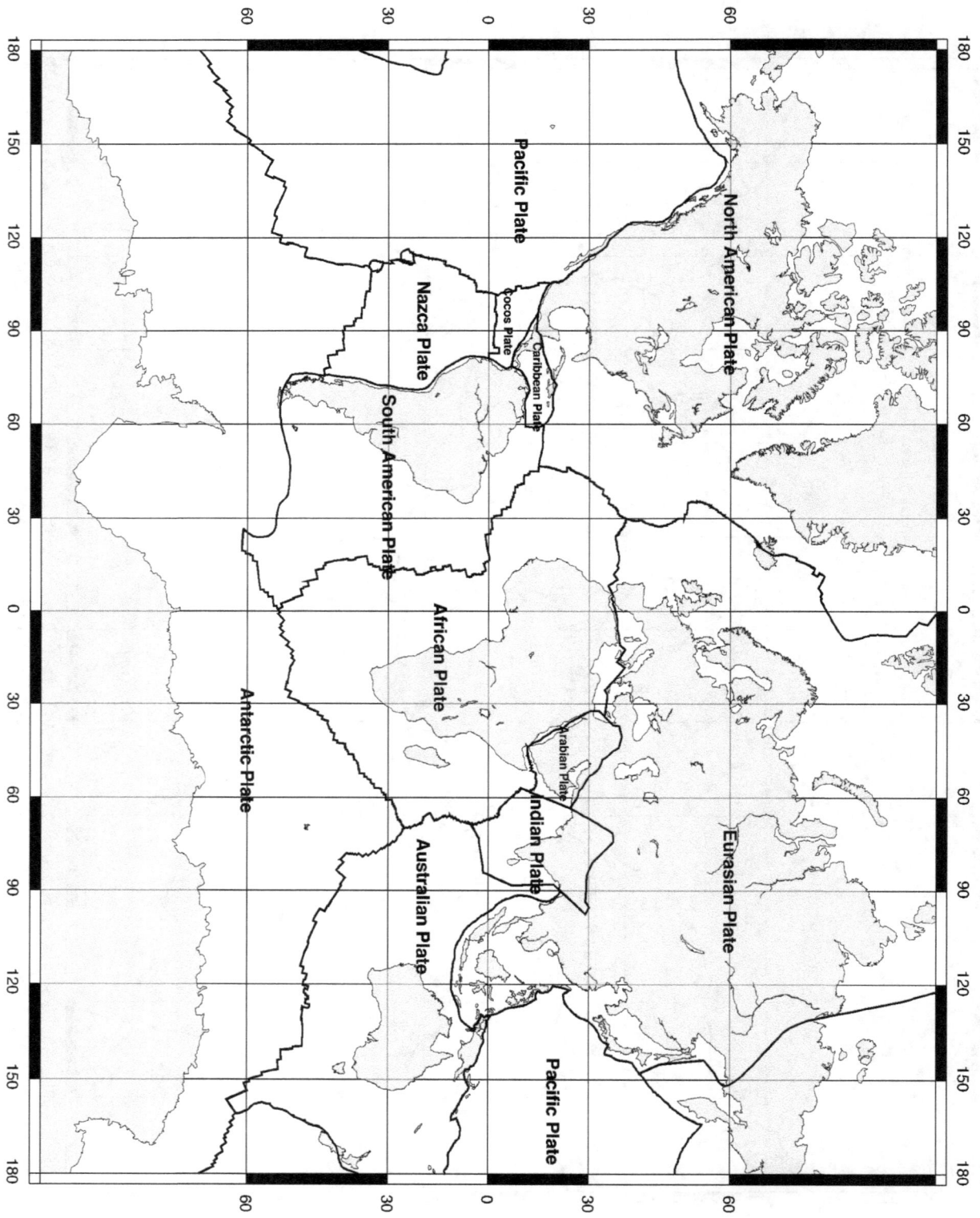

Figure 3: Plate boundary map. This map is used courtesy of Dale S. Sawyer and Rice University (http://plateboundary.rice.edu/downloads.html).

SCIENTIFIC SPECIALTY: SEISMOLOGY

Earthquake Locations 1990 - 1996 (Magnitudes 4 and greater)

Color indicates depth: Red 0-33 km, Orange 33-70 km, Green 70-300 km, Blue 300-700 km

This map is part of "Discovering Plate Boundaries," a classroom exercise developed by Dale S. Sawyer at Rice University (dale@rice.edu). Additional information about this exercise can be found at http://terra.rice.edu/plateboundary .

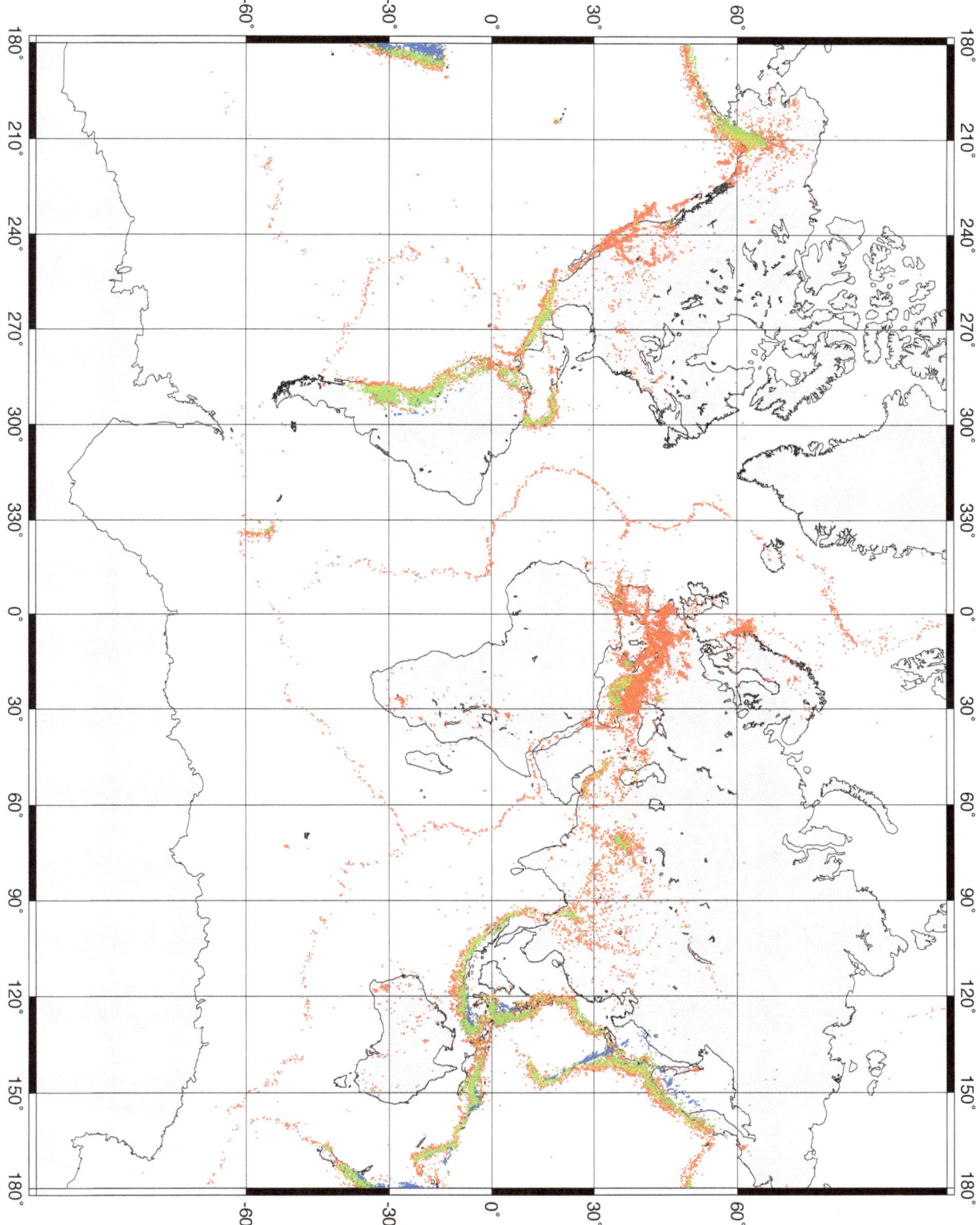

Figure 4: Earthquake depth and distribution map. This map is used courtesy of Dale S. Sawyer and Rice University (http://plateboundary.rice.edu/downloads.html).

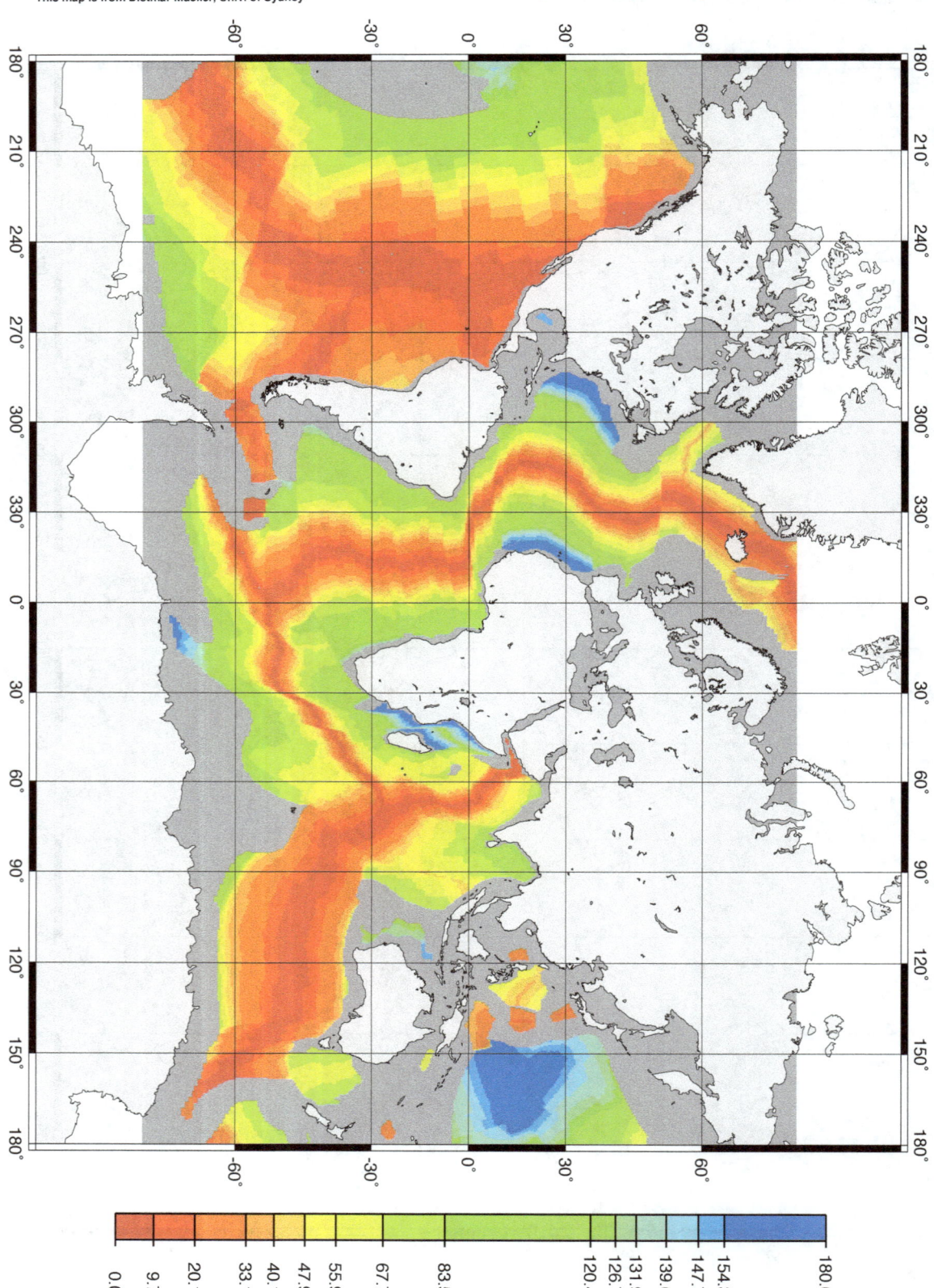

Figure 5: Age of the ocean floor. This map is used courtesy of Dale S. Sawyer and Rice University (http://plateboundary.rice.edu/downloads.html).

Lab 4
Mineral Properties

Name _____   Date _____

Purpose: To learn about the various properties used to identify minerals.
Materials: Mineral kits for hardness, cleavage, and luster
Instructions: Using the provided kits fill in the tables below.

PART 1: MINERAL HARDNESS

DEFINITIONS: **Hardness** is a measure of how well a mineral resists being scratched. It is related to the strength of the bonds within a mineral. The hardness of minerals is expressed on the Mohs Hardness Scale from 1 (the softest mineral) to 10 (the hardest). The relative hardness of a mineral can be established by testing it against other minerals and against objects of known hardness. The objects commonly used are:

- Fingernail.................... 2 ½
- Copper penny .... 3 ½
- Wire nail .... 4 ½
- Glass............................. 5 ½
- Porcelain streak plate...... 6 ½

PROCEDURE: Using the provided hardness kit complete the table below (Table 1). To determine hardness start by scratching the sample with your fingernail (you can either scratch the sample with your fingernail or scratch your fingernail with the sample).

    a. If the sample is scratched it is softer than 2.5
    b. if your fingernail is scratched the sample is harder than 2.5
    c. if both your fingernail and sample are scratched the sample has a hardness of 2.5

If the sample is harder than 2.5, move on to the penny. Continue until you find the hardness of the mineral.

Table 1: Mineral Hardness

Sample Number	Hardness

PART 2: COLOR, LUSTER AND STREAK

DEFINITIONS:

a. **Color** is an obvious property of minerals. The color of some minerals is consistent from sample to sample and is useful for identification. Many minerals occur in a variety of colors, however, so you cannot rely on it for identification.

b. **Streak** is the color of the powder of a mineral. Streak is more consistent and useful than color for identifying many minerals, especially metallic ones.

c. **Luster** describes how a mineral reflects light back to the eye. Luster is useful in identification of many minerals, though some minerals can display more than one type of luster, and more than one luster may be observed in a single sample.

PROCEDURE: Use the provided color, luster and streak kit(s) to fill in the following table (Table 2).

1. Record the overall color of the mineral
2. To determine the streak rub the mineral along the porcelain streak plate. Do not push very hard, try to avoid breaking pieces of the sample off. Be specific about the color of the streak. If there is no streak put "none".
3. The primary description for luster is metallic or non-metallic. If available it is best to use a fresh surface to identify luster. A sample with metallic luster shines like a metal, meaning they are reflective and opaque (does not allow light through). Not all shiny samples are metallic, so be careful.
4. If the sample is non-metallic use the list below to further describe it.
    - Vitreous - Glassy luster
    - Adamantine - Diamond-like brilliance
    - Resinous - Looks like sap or honey
    - Silky - Cloth-like reflection
    - Pearly - Shines like a pearl
    - Greasy - An appearance like oil or grease
    - Waxy - Reflects light like candle wax
    - Dull or Earthy - no shine at all

Table 2: Color, Luster, and Streak

Sample	Color	Streak	Luster

PART 3: CLEAVAGE AND FRACTURE

DEFINITIONS: **Cleavage** is a property some minerals have where they break easily along certain planes. These planes of weakness are related to the mineral structure. When broken they tend to be flat and shiny. Because mineral structure repeats through a crystal, cleavage planes repeat through a crystal as well, leading to a stair-step pattern of flat surfaces in a crystal. Some minerals have multiple cleavage planes oriented in different directions. When this occurs we describe the cleavage pattern based on the number of distinct planes and the angles between those planes. The common cleavage patterns are as follows (refer to mineral properties discussion in your textbook for images):

- One direction: The mineral breaks in sheets (like a deck of cards). Basal cleavage.
- Two directions at ninety degrees: Two planes that meet at a right angle - creating a square prism.
- Two directions not at ninety degrees: Two planes that do not meet at right angles - forming a non-rectangular prism.
- Three directions at ninety degrees: Three planes, all meeting at right angles - cubic cleavage.
- Three directions not at ninety degrees: Three planes not meeting at right angles - rhombic cleavage (sort of a tilted cube).
- Four directions: Harder to picture, in ideal samples it creates an octahedron.
- Six directions: Very hard to imagine, much harder to identify - creates a dodecahedron.

To help you decide if a mineral has cleavage, rotate it under a good light source and observe whether there is some position in which the surface of the specimen appears to "light up" (like the flash of light across a mirror or watch face). Cleavage planes reflect more light than other breakage surfaces and thus appear shinier. Some minerals exhibit "perfect" or "excellent" cleavage with obvious large, shiny, flat surfaces and sometimes with very characteristic shapes. Others exhibit cleavage that is considered "good" or "poor" which are less obvious.

Not all minerals have cleavage. When a crystal structure does not have any weak planes it will break unevenly. This is known as **fracture**. There are several types of fracture:

- Conchoidal: A fracture pattern of concentric circles, forming a shell-like pattern.
- Fibrous or Splintery: Parallel crystals that split apart from each other.
- Earthy: Smooth (but dull) breakage like clay or chalk
- Uneven or Irregular: Rough breakages with no distinct pattern.

PROCEDURE: Using the provided cleavage and fracture kit, fill in the following table (Table 3). Do not actually break the samples.

Table 3: Cleavage and Fracture

Sample	Does the Minerals have Cleavage? Yes or No	If the sample has cleavage: How many directions?	If more than one direction: Are they at 90 degrees or not at 90 degrees?	If the sample has no cleavage: Describe the fracture

PART 4: OTHER PROPERTIES

DEFINITIONS:

a. **Density**: an objects mass per unit of volume, or in other words, how much it weighs compared to how much space it takes up. It is represented by the following formula where mass is in grams (g) and volume is in cubic centimeters (cm3):

b. **Specific Gravity**: Geologists typically use a related measure, specific gravity, to describe the density or "heft" of a mineral or rock. Specific gravity (SG) is a ratio of the density of a substance (such as a mineral) to the density of water (1.0 g/cm3).

c. **Reaction to Acid**: Certain minerals, such as carbonates will react with acid by effervescing or "fizzing" when a small amount of dilute acid (typically hydrochloric acid, HCl) is applied.

d. **Magnetism**: Many minerals contain the element Iron (Fe). If a mineral has a high enough Fe concentration, it will attract a magnet. To test this, simply hold a hand magnet close to the sample and see if you feel a pull.

e. **Double Refraction** is an optical property present in very few minerals where an image passing through a transparent piece of the mineral appears to be doubled. It occurs when a light beam is split into two perpendicular directions due to the arrangement of the atoms.

f. **Striations** are fine, parallel grooves that may be present on the crystal face or cleavage plane of some minerals. Quartz crystals and plagioclase feldspars exhibit striations.

g. **Taste**: Some minerals have a distinct taste. For example, halite (NaCl) is identical to table salt (and therefore tastes like it!) and sylvite (KCl), which is often used as a salt substitute for those watching their sodium (Na) intake, has a more bitter taste. Kaolinite, a clay mineral, tastes of aluminum. Do not taste a mineral without asking your instructor first.

h. **Feel**: You will find that several minerals have a distinct feel. Talc, for example, has a soapy feel (like a bar of soap) while graphite feels greasy.

i. **Odor**: Certain minerals give off a distinct smell. This is especially true if a mineral contains the element sulfur (S). Galena (PbS) and sphalerite (ZnS) both smell strongly of sulfur (like rotting eggs, a match head, or sometimes well water). Kaolinite will give off a distinct "earthy" odor when breathed upon first.

Lab 5
Rock Forming Minerals

Name _____   Date _____

Purpose: Describe and identify common rock forming minerals.
Materials:

- Rock Forming Minerals kit
- Hardness kit
- Hand lens
- Dilute HCl

Instructions: Fill in Table 1. Use the provided identification tables to name the mineral.

Table 1: Types of Rock Forming Minerals

ID #	Luster	Color of Streak	Color of mineral	Cleavage yes/no	Cleavage pattern	Hardness	Other	Mineral Name

Lab 6
Economic Minerals

Name _____ Date _____

Purpose: Describe and identify common economic minerals.

Materials:
- Economic Minerals kit
- Hardness kit
- Hand lens
- Dilute HCl

Instructions: Fill in Table 1. Use the provided identification tables to name the mineral.

Table 1: Types of Economic Minerals

ID #	Luster	Color of Streak	Color of mineral	Cleavage yes/no	Cleavage pattern	Hardness	Other	Mineral Name

Metallic Luster

H < 5½

Hardness	Color	Streak	Luster	Cleavage/Fracture	Other Properties	Name and Chemical Composition
1.0-2.0	lead-gray to black	heavy gray	metallic to dull	basal cleavage, easily weathers into fine flakes	low SG (2.1) for mineral with metallic luster; will write on paper; greasy feel	GRAPHITE; C
2.5	bright lead gray	heavy gray	metallic	excellent cubic cleavage	sulfur smell; high SG (7.5)	GALENA; PbS
2.5-3.0	brassy-yellow	yellow-gold	metallic	hackly or irregular fracture	very high SG (17.2)	GOLD; Au
2.5-3.0	silver-steel	silver-white	metallic	hackly or irregular fracture	high SG (10.3)	SILVER; Ag
2.5-3.0	copper (penny)	copper-red	metallic	hackly or irregular fracture	high SG (8.8); often has green discoloration from weathering	COPPER; Cu
3.0	brassy-brown, mottlings of blue or purple	gray-black	metallic	irregular fracture	commonly massive, micro-crystalline	BORNITE; Cu$_5$FeS$_4$
3.5	brassy-yellow	black	metallic	irregular fracture	commonly massive	CHALCOPYRITE; CuFeS$_2$
3.5-4.0	brown to orange to red	yellow	highly vitreous	dodecahedral – 6 directions!	Faint sulfur smell; SG = 4.0	SPHALERITE; ZnS
1.5-5.5	orange, brown, or yellow	yellow	sub-metallic & earthy varieties	irregular fracture	non-crystalline (amorphous)	LIMONITE; Fe$_2$O$_3$nH$_2$O (see also non-metallic)
5.0-6.0	black	black	dull to sub-metallic	conchoidal fracture	non-crystalline (amorphous)	PSILOMELANE; BaMn$_9$O$_{16}$(OH)$_4$

Metallic Luster

H > 5½

Hardness	Color	Streak	Luster	Cleavage/Fracture	Other Properties	Name and Chemical Composition
5.5-6.0	silver-gray-white	black	metallic	prismatic cleavage	bitter smell on fresh surfaces	ARSENOPYRITE; FeAsS
6.0-6.5	brass-yellow	greenish-black	metallic	irregular fracture	"Fool's Gold"; often forms as cubes or pyritohedrons	PYRITE; FeS$_2$
5.5-6.5	steel-gray, dull black, or deep, earthy red, depending on variety	red	metallic	irregular fracture	three common varieties: specular (bright metallic), oolitic (containing small, spherical masses), sedimentary (earthy and usually deep red). Hardness can be difficult to determine since some specimens easily crumble.	HEMATITE; Fe$_2$O$_3$ (see also non-metallic)
6.0	dark gray to black	black	metallic to dull black (if weathered)	irregular fracture	can be strongly magnetic	MAGNETITE; Fe$_3$O$_4$

Non Metallic Luster

H < 2½

Hardness	Color	Streak	Luster	Cleavage/Fracture	Other Properties	Name and Chemical Composition
1.0	white, gray, light yellow, pink, blue or light green	white	pearly, silky	basal cleavage, but weathers easily	very soft; soapy feel	TALC; $Mg_3Si_4O_{10}(OH)_2$
2.0	clear and colorless to frosty white	white	vitreous to dull	basal cleavage	leaves a frosty appearance	GYPSUM; $CaSO_4 \cdot 2H_2O$
2.0	colorless, white, or with shades of red or yellow	white	vitreous	perfect cubic cleavage	salty and bitter taste	SYLVITE; KCl
2.0-2.5	white, gray-brown	white	vitreous, transparent in thin sections	perfect basal cleavage	part of mica group; thin, flexible sheets or "books"	MUSCOVITE (Mica); $KAl_2(AlSi_3O_{10})(OH)_2$
2.0-2.5	dark green	faint green to yellow	pearly or dull	basal cleavage	forms short prisms that split into flexible sheets	CHLORITE; hydrous K-Mg-Al silicate
1.5-2.5	bright yellow	white to yellow	usually earthy and opaque	irregular fracture	strong odor	SULFUR; S
2.0-2.5	white	white	earthy	irregular fracture	easily leaves powder behind on surfaces; earthy-type smell when damp	KAOLINITE; $Al_2Si_2O_5(OH)_4$
1.5-5.5 (variable)	orange, brown, or yellow	yellow, orange, or reddish	sub-metallic & earthy varieties	irregular fracture	non-crystalline (amorphous)	LIMONITE; $Fe_2O_3 \cdot nH_2O$ (see also metallic)

Non Metallic Luster

H 2½ - 5½ WITH CLEAVAGE

Hardness	Color	Streak	Luster	Cleavage/ Fracture	Other Properties	Name and Chemical Composition
2.5	colorless to white	white	glassy	perfect cubic cleavage	tastes like salt; soluble in water, transparent to translucent	HALITE; NaCl
2.5-3	dark brown to black	slight greenish	glassy or splendent	perfect basal cleavage	part of mica group; thin, flexible sheets or "books"	BIOTITE (Mica); K(Mg,Fe)$_3$(AlSi$_3$O$_{10}$)(OH)$_2$
3.5-4.0	usually a shade of pink, but can be white, brown	white	vitreous to pearly	perfect rhombic	effervesces weakly in acid; usually small, elongate crystals with saddle-like faces	DOLOMITE; CaMg(CO$_3$)$_2$
3.5-4.0	brown to orange to red	yellow	highly vitreous	dodecahedral – 6 directions!	Faint sulfur smell; SG = 4.0	SPHALERITE; ZnS
3.0	usually colorless to white, can be orange, gray, yellow, or brown	white	glassy	excellent rhombic cleavage	strongly reacts to dilute acid	CALCITE; CaCO$_3$
3-3.5	colorless to white (often with tints of brown, yellow, or red)	white	glassy or pearly	excellent cleavage	high SG for non-metallic (4.5); forms short tabular crystals or rose shaped masses	BARITE; BaSO$_4$
4.0	almost any color (colorless, light green, blue, yellow, purple)	white	glassy	octahedral	Exhibits fluorescence, may have slight greasy feel	FLUORITE; CaF$_2$
5.0-6.0	dark green to black	white	vitreous	two cleavage directions at about 60° & 120°	forms elongated crystals; sometimes bladed; most common amphibole	HORNBLENDE; CaMg$_3$(SiO$_3$)$_4$
5.0-6.0	dark green to black	white	vitreous	two cleavage directions at about 90°	most common pyroxene	AUGITE; (very complex silicate)

Non Metallic Luster

H 2½ - 5½ WITH CLEAVAGE

Hardness	Color	Streak	Luster	Cleavage/Fracture	Other Properties	Name and Chemical Composition
3.5-4.0	azure blue	blue	dull to vitreous	irregular fracture	crusts, laminated masses, or prisms; reacts slowly with acid; usually found with malachite	AZURITE; $Cu_3CO_3(OH)_2$
3.5-4.0	bright green (Paris green)	green	dull to vitreous	irregular fracture	crusts, laminated masses, or prisms; reacts slowly with acid; usually found with azurite	MALACHITE; $Cu_2CO_3(OH)_2$
2.5-3.5	white, gray, red, rust color	light	dull, earthy	irregular fracture	contains rounded masses with an earthy matrix	BAUXITE; hydrous oxides of aluminum
5.0	green, yellow, variable	white	vitreous to sub-resinous	conchoidal fracture	hexagonal crystals	APATITE; $Ca_5(PO_4)_3(OH,F,Cl)$
2.5-5.0 (varies)	light to dark green, sometimes with gray	white	waxy, silky	conchoidal/splintery	some fibrous varieties (asbestos)	SERPENTINE; $Mg_6(Si_4O_{10})(OH)_8$
5.5-6.5	steel-gray, dull black, or deep, earthy red, depending on variety	red	metallic	irregular fracture	three common varieties: specular (bright metallic), oolitic (containing small, spherical masses), sedimentary (earthy and usually deep red). Hardness can be difficult to determine since some specimens easily crumble."	HEMATITE; Fe_2O_3 (see also non-metallic)

Non Metallic Luster

H 2½ - 5½

Hardness	Color	Streak	Luster	Cleavage/Fracture	Other Properties	Name and Chemical Composition
5.0-6.0	dark green to black	white	vitreous	two cleavage directions at about 60° & 120°	forms elongated crystals; sometimes bladed; most common amphibole	HORNBLENDE; CaMg₃(SiO₃)₄
5.0-6.0	dark green to black	white	vitreous	two cleavage directions at about 90°	most common pyroxene	AUGITE; (very complex silicate)
6.0	usually varying shades of pink ("salmon" pink) sometimes white	white	vitreous to pearly	two directions nearly at 90°	thin veinlets on cleavage planes: variety of Potassium Feldspar	ORTHOCLASE; KAlSi₃O₈
6.0	black to blue-black	white	high vitreous	two directions nearly at 90°	thin veinlets on cleavage planes; will show iridescence	LABRADORITE; (Ca,Na)(Al,Si)AlSi₂O₈ (Plagioclase Feldspar)
6.0	white, gray, green	white	vitreous to pearly	two directions nearly at 90°	striations on some cleavage planes	ALBITE; NaAlSi₃O₈ (Plagioclase Feldspar)
4.5-6.5	blue to gray-blue	white	vitreous	perfect basal	bladed crystals; hardness varies - 4.5 (long axis), 6.5 (short axis)	KYANITE; Al₂SiO₅
6.5-7.0	light to dark green		vitreous	conchoidal fracture	often a mass of small crystals, with an appearance like granulated sugar	OLIVINE; (MgFe)₂SiO₄
7.0	colorless, white, purple, pink, yellow, brown		vitreous	conchoidal fracture	many varieties (e.g., smoky, milky, citrine, rose, amethyst, tiger's eye, etc.)	QUARTZ; SiO₂
7.0	can be white, gray, blue, brown, red	white	vitreous to sub-vitreous to dull	conchoidal fracture	many varieties (e.g., jasper, flint, chalcedony, agate, etc.), can form very sharp edges	MICROCRYSTALLINE QUARTZ; SiO₂
7.0-7.5	deep red to reddish-brown		vitreous to resinous	sub-conchoidal fracture	usually occurs as perfect crystals	GARNET; complex silicate
7.5	red-brown to brown-black		vitreous to resinous	sub-conchoidal fracture	usually twinned at 90° (cruciform) or 60°	STAUROLITE; FeAl₄Si₂O₁₀(OH)₂
8.0	colorless, yellow, brown, etc		glassy to opaque	1 directional	transparent to opaque	TOPAZ; (AlF)₂SiO₄
9.0	brown to red, sometimes bluish gray		adamantine to opaque	often 1 direction of "false cleavage"	very often hexagonal; SG = 4.0; rubies and sapphires are gem-quality corundum	CORUNDUM; Al₂O₃

Lab 7
M&M Magma Chamber

Name _____ Date _____

Purpose: To understand Bowen's Reaction Series and the process of fractional crystallization within a simulated magma chamber.

Materials:

- M&M's separated by color or artificial replacement
- Large piece of white paper or poster board
- Calculator

Instructions:

1. Determine the proportions of atoms in each mineral. Record these on the data sheet in the section of the attached table labeled Mineral Compositions.
2. Wash your hands.
3. Sort the M&Ms by color (if not already done). Carefully count the number of M&Ms that represent each element in the starting magma composition in the table on Mineral Compositions.
4. Mix well and place at one end of the white sheet of paper to form your magma chamber.
5. Begin crystallization of magma by removing appropriate numbers of M&Ms at each crystallization step. Minerals (M&Ms) that are crystallized in each step are moved to the other end of the magma chamber.
6. After each crystallization step, tally the number of atoms remaining for each of the elements (colors of M&Ms). Record on the M&M's Remaining table.
7. Continue with the next crystallization step.
8. For each crystallization step, calculate the relative proportions of each element (colors of M&M) remaining as a percentage of the total remaining after that crystallization step. Record this information in the Magma Composition table.
9. Also calculate the proportion of magma remaining as a fraction of the original number of M&Ms. Record this information in the Magma Composition table.
10. Plot the percentage of each cation (M&M color) in the magma on the attached graph paper — or use an Excel spreadsheet to create your own plots.
11. Answer the follow-up questions.

MINERAL COMPOSITIONS

Color	Initial	Mg-Ol	Fe-Ol	Pyx	Ca-Plag	Na-Plag	Magnetite	Quartz
	Magma	Mg$_2$SiO$_4$	Fe$_2$SiO$_4$	CaMgSi$_2$O$_6$	CaAl$_2$Si$_2$O$_8$	NaAlSi$_3$O$_8$	Fe$_3$O$_4$	SiO$_2$
Si	184							
Al	70							
Fe	39							
Mg	40							
Ca	34							
Na	24							
Total	391							

MINERALS CRYSTALLIZED

Minerals	Step 1	Step 2	Step 3	Step 4	Step 5	Step 6	Step 7	Step 8	Step 9	Step 10
Mg-Ol	2	3	4	3	2					
Fe-Ol			1	1	1	1				
Pyx				1	1	4	2	2	1	1
Ca-Plag			2	3	3	4	3	3	1	1
Na-Plag				1	1	3	3	6	5	7
Quartz										4
Magnetite						1	2	2	1	1
Total	2	4	7	8	8	13	10	13	9	14

M&M'S REMAINING

Elements	Initial	1	2	3	4	5	6	7	8	9	10
Si	184										
Al	70										
Fe	39										
Mg	40										
Ca	34										
Na	24										
Total	391										

MAGMA COMPOSITION

Elements	Initial	1	2	3	4	5	6	7	8	9	10
Si	47.1										
Al	17.9										
Fe	10.0										
Mg	10.2										
Ca	8.7										
Na	6.1										
Total	100.00										

1. Compare the types of minerals removed at the beginning of crystallization with those removed in the middle and end of crystallization.

2. Apply an igneous rock name to the minerals that were removed in during the 2nd, 6th, and 10th crystallization steps. Use the phaneritic (coarse grained) rock names (use your textbook for igneous identification charts in Chapter 4).

Step	Rock Name
2nd	
6th	
10th	

3. Compare the trends of the different elements throughout crystallization. Refer to your graphs in your answer.

4. Magmas are broadly classified by their silica composition according to the following scheme:

   Si%   40   50   60   70
              Basalt   Andesite   Rhyolite

   Using the above classification scheme, classify the magma at the following steps:

Step	Rock Name
	Initial magma
2nd	
6th	
10th	

5. Explain how the percentage of silica in magma increases during crystallization despite the fact that silicate minerals are being removed throughout the crystallization process.

6. Which aspects of this model magma chamber are realistic? Which are not? Suggest some ways to make the model more realistic.

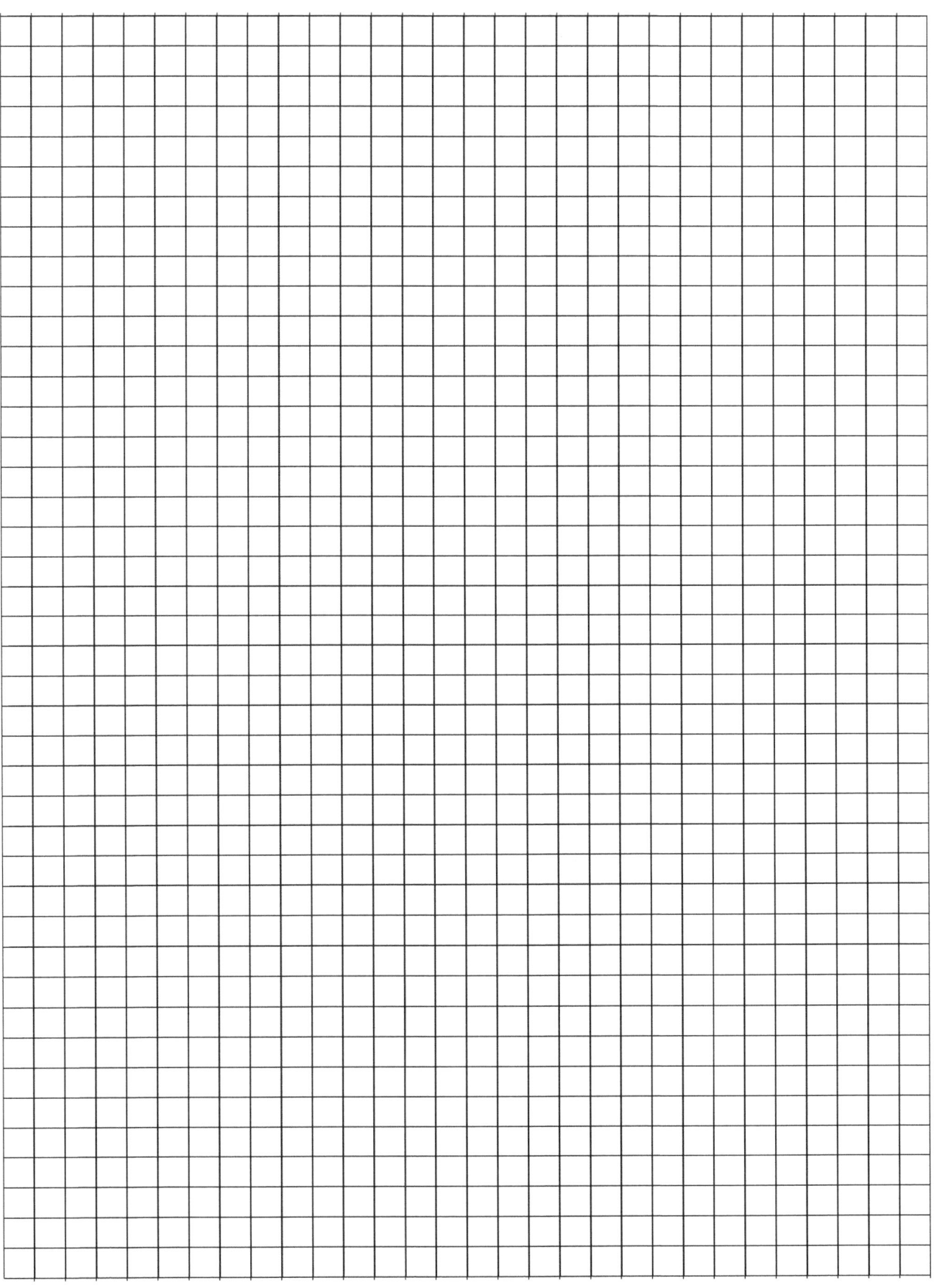

Lab 8
Igneous Rocks

Name _____ Date _____

Purpose: Describe and identify common igneous rocks.
Materials:

- Igneous rock samples
- Hand lens
- Glass plate (optional)

Instructions: Complete the identification table (Table 2) using the following criteria for each column.

TEXTURE

Textures of igneous rocks are based on the size, shape, and arrangement of crystals within the rock.

- **Aphanitic** (fine-grained): Mineral crystals in the rock are too small to see with your eye.
- **Phaneritic** (coarse-grained): All mineral crystals are visible to the eye and a clear distinction can be made between different mineral grains (even if they are the same color).
- **Porphyritic**: Two distinct grain sizes (like a chocolate chip cookie) You will see larger crystals embedded in a fine-grained (aphanitic) "background" material. Think of it as a phaneritic/aphanitic "hybrid".
- **Glassy**: No trick, it looks like glass. There are no crystal grains do to really rapid cooling.
- **Vesicular**: A spongy, porous look to the sample caused by trapped gas bubbles in the rock during cooling process.
- **Pyroclastic**: Made from exploded rock fragments and ash fused together in the heat of an eruption.
- **Pegmatite**: Interlocking crystals usually larger than 2.5 cm in size (1 in)

ORIGIN

The origin of the igneous rock is based on where the magma cooled/crystallized and is determined by texture designation.

- **Volcanic** (extrusive): Rock cooled/formed above ground (result of lava flow or eruption).
- **Plutonic** (intrusive): Rock cooled/formed underground (igneous intrusion).

COLOR

Relative or overall color is related to the percentage of dark-colored minerals. Igneous rocks can be light-colored (lots of white and pink), dark-colored (dark green and black), or "medium"-colored (some light/some dark minerals, or an overall gray coloring).

COMPOSITION

Color is a good indicator of igneous composition. If the sample is aphanitic use the color to estimate composition (remember though - color is not always reliable). If the sample is phaneritic or porphyritic use your hand lens to identify the minerals within the rock.

- **Felsic**: Light-colored igneous rocks tend to be felsic, containing mostly light silicates and having a relatively high silica content.
- **Mafic**: Dark-colored igneous rocks tend to be mafic, containing mostly dark silicates and having a relatively low silica content.
- **Intermediate**: Gray or "medium" colored igneous rocks tend to be intermediate – between felsic and mafic.
- **Ultramafic**: Ultramafic igneous rocks are less common. They have very high percentages of dark silicates and low percentages of silica. They will typically be quite dark black and dark green, but can also be a light "olive" green color.

Rock Name: Use Table 1 to identify the rock.

Table 1: Types of Igneous Rocks

	Felsic	Intermediate	Mafic	Ultramafic
Phaneritic	Granite (If REALLY coarse grained: pegmatite)	Diorite	Gabbro	Peridotite (if mostly olivine: dunite)
Aphanitic	Rhyolite	Andesite	Basalt	Komatiite
Porphyritic				
Vesicular	Pumice		Scoria	
Glassy	Obsidian			
Pyroclastic	Tuff or Volcanic Breccia			

Table 2: Identifying Igneous Rocks

ID #	Texture	Volcanic or Plutonic	Color	Composition or Mineralogy	Rock Name	Other Characteristics

Lab 9
Igneous Rocks of Oregon

Name _____ Date _____

Purpose: Explore the plutonic and volcanic rocks of Oregon.
Materials: Geologic Map of Oregon
Instructions: Use the provided map to answer the following questions. When asked for a rock unit or formation provide the symbol and name from the key.

1. What rock units are found around Cascade Volcanoes like Hood and Jefferson? What does the age and rock type tell you about the volcanoes' eruptive style and history?

2. Which is older, the Western or High Cascades (a map of Oregon provinces is provided on the key of the geologic map).

3. How are the eruption styles between the the Western and High Cascades different? Use evidence from the map (what rock units are present).

4. An **isopach** is a line on a geologic map that connects areas of equal thickness of a material. Ash isopachs from Mt. Mazama and Newbery volcano eruptions are shown.

    a. Estimate the thickness of the Mazama ash that likely exists at the botom of Odell Lake (roughly in the middle).

    b. Ash from the Newberry eruption thins in what direction?

    c. In what direction does the wind more or less constantly move for the last few thousand years at least? What is the most abundant igneous rock unit found on the map? Where is it exposed? Where did it come from?

5. What is the main (most abundant) plutonic rock found on the map? Where is it?

6. Can you find any obsidian on the map? Why wouldn't you find a deposit large enough to be on the map?

7. Which is older, the volcanic or plutonic rocks of Oregon? How do you know this?

8. What volcanic rocks are found in the Coast Range?

9. What types of plutonic rocks do you find in the Coast Range? How are they similar to or different from the volcanic rocks of the same location?

10. Are the Coast Range mountains volcanoes? Why or why not?

11. Compare the volcanic rocks of NE Oregon to the rocks of SE Oregon, how are they different?

12. What do the rock types suggest about the recent volcanic activity in SE Oregon compared to NE Oregon?

13. This is a surface geology map, meaning it only shows what sediment or rock unit is found at the surface of each location. Many areas of the Willamette valley are covered by Qal, Qs or similar units. What are these?

14. Provide an explanation for the lack of quaternary units like Qal or Qs in Eastern Oregon (i.e., why do so many areas have exposed rock at the surface instead of unconsolidated sediments)?

Lab 10
Sedimentary Rocks

Name _____ Date _____

Purpose: Describe and identify common sedimentary rocks.
Materials:

- Sedimentary rock samples
- Dilute HCl
- Hand lens
- Glass plate (optional)

Instructions: Complete the identification table on page 3 using the following criteria for each column.

SEDIMENT SOURCE

Sediment source for sedimentary rocks is essentially the way sediment accumulated to form the rock.

- **Detrital**: originates and is transported as solid particles derived from both mechanical and chemical weathering.
- **Chemical**: originates as soluble material produced primarily by chemical weathering. Ions are transported in solution (water) and are then precipitated by inorganic or biological processes.
- **Organic**: composed of organic material. The "sediment" of organic sedimentary rocks is the carbon-rich remains of organisms (partially decayed plant material) and not minerals derived from weathering.

TEXTURE

Textures of sedimentary rocks are based on the particle size or arrangement of crystals within the rock.

- **Clastic**: weathered pieces of rock or mineral fragments cemented together
- **Crystalline**: interlocking crystals

Note: all detrital rocks are clastic, but chemical rocks can be clastic or crystalline.

GRAIN SIZE

Grain size only applies to detrital rocks! It refers is the predominant sediment size that makes up the rock. Options are: clay, silt, sand, and gravel. (See sedimentary classification chart below for grain diameters.)

COMPOSITION

Composition of sedimentary rock refers to its mineral make-up. Some sedimentary rocks are defined by their specific mineral compositions (e.g., limestone, chert) while others are composed rock fragments with a variety of minerals. Some common sedimentary minerals include quartz (silica), calcite, gypsum, halite, and potassium feldspar.

Rock Name: Use Table 1 on the following page to identify the rock.

Table 1: Sedimentary Classifications

Detrital Rocks			Chemical, Biochemical and Organic Rocks		
Clastic Texture	Grain diameter	Rock Name	Composition	Texture	Rock name
Gravel - Round Pieces	> 2mm	Conglomerate	Calcite	Crystalline	Crystalline Limestone
Gravel - Angular Pieces	> 2mm	Breccia	Calcite	Clastic - sand sized grains	Oolitic Limestone
Sand - Quartz	.06mm–2mm Visible	Quartz Sandstone	Calcite	Crystalline - often porous	Travertine
Sand - Quartz and Feldspar	.06mm–2mm Visible	Arkose	Calcite	Clastic - microscopic shells	Chalk
Sand - Rock Fragments	.06 mm–2 mm Visible	Graywacke	Calcite	Clastic - shells	Fossiliferous Limestone
Silt	.004mm–.06mm Not visible	Siltstone	Calcite	Clastic - entirely shells	Coquina
Clay	<.004 mm Not visible	Shale	Quartz	Crystalline (very fine)	Chert or Flint
			Gypsum	Crystalline	Rock Gypsum
			Halite	Crystalline	Rock Salt
			Organic Material	Non-clastic	Coal
			Dolomite	Crystalline	Dolostone

Table 2: Identifying Sedimentary Rocks

ID #	Sediment Source	Texture	Grain Size (Detrital only)	Mineral Composition	Rock Name	Other Characteristics

Lab 11
Sedimentary Rocks of Oregon

Name _____   Date _____

Purpose: Explore the sedimentary rocks of Oregon.
Materials:

- Geologic Map of Oregon
- Sedimentary Rock Kit
- Hand lens

Instructions:

PART 1

a. Fill in the following table using the samples available to you.
b. For the detrital rocks, use the second column to describe the shape and sorting of the grains
c. For the non-detrital rocks, use the second column to list any distinguishing features in the rock that may indicate the environment the rock formed in.

Table 1: Identifying Sedimentary Rocks of Oregon

Sample	Grain Shape & Sorting or Distinguishing Features	Environment What conditions or environment most likely created this rock (e.g., desert, stream, marine, coast etc.)

PART 2

Use the provided map to answer the following questions. When asked for a rock unit or formation provide the symbol and name.

1. Where are the oldest sedimentary rocks in Oregon? What kind of rocks are they? What environment did they form in?

2. Which unit symbol represents the most recent deposits (which time period or epoch)?

3. What is Qal? Where is it always found?

4. What is Qd? Where is it found (two locations)? Why would the same features appear in such drastically different environments?

5. What is Qs? If I told you the deposits in the Willamette Valley were 13,000 to 15000 years old, what do you think lacustrine means (Hint: Missoula)?

6. Given your answer to number 6, what do the lacustrine deposits in SE Oregon indicate about the past environments there?

7. What is Tt? Where is it found? These were formed by turbidity currents. What environment did the sediments initially form in?

8. If Tt now makes up several mountain peaks, what does that mean about the tectonic history of the area?

9. Look through the rock descriptions (you actually have to go through the descriptions in detail, do not just read the formation names). Are there any units with coal deposits in them? What does this suggest about their environments?

10. Notice many sedimentary units along the coast contain arkosic and tuffaceous deposits. What does this suggest about the environment and sources of the sediments that make up the rocks (i.e., what do arkosic and tuffaceous mean)?

11. What types of sedimentary rocks or deposits are found in the Cascades (Western or High Cascades)? Provide some examples.

12. What does this collection of sediment types suggest about the primary *sedimentary environment(s)* seen in the Cascades over the history of Oregon?

13. Where in the state do you generally find fanglomerate? Given the description in the key and its locations on the map, what conditions does the rock most likely represent (for example, long transport or short, dry or wet environments)?

14. Which units in the Blue Mountains and Klamath Mountains suggest the two share a common history (ignoring Quaternary deposits)?

15. Central Oregon is currently a mid-latitude desert as a result of the rain shadow created by the Cascades. But what type of environment is represented in the rocks from the Mesozoic Era? List two or three of the sedimentary rock units in Central Oregon that support this.

Lab 12
Metamorphic Rocks

Name _____   Date _____

Purpose: Describe and identify common metamorphic rocks.
Materials:

- Metamorphic rock samples
- Glass Plate
- Dilute HCl
- Hand Lens

Instructions: Fill in the identification table (Table 2). Use the following criteria for each column.

TEXTURE

Textures of metamorphic rocks are based on the shape and arrangement of crystals within the rock.

- **Foliated**: Rock exhibits foliation (aligned minerals by pressure or "rock cleavage")
- **Non-foliated**: Does not have foliation

CRYSTAL SIZE

- **Coarse-grained**: crystals are visible
- **Fine-grained**: crystals are not visible

MINERALOGY

What minerals are apparent in the rock? Use your knowledge of minerals here (see Mineral Tables if necessary). Some helpful tests will include:

- Reaction to acid: This is only necessary for non-foliated rocks. Place one drop of dilute HCl on to the rock.
- Glass Scratch Test: This is only necessary for non-foliated rocks. (Reminder: Mohs hardness of glass = 5.5)

Rock Name: Use Table 1 to identify the rock.

Table 1: Types of Metamorphic Rocks

Foliation	Features/minerals	Rock Name
Foliated	banded	gneiss
	coarse grained	Schist (If blue - blueschist)
	fine grained, reflective	phyllite
	fine grained	slate
maybe foliated	garnet and pyroxene common	eclogite
non-foliated	It's green (chlorite, actinolite, epidote)	greenstone
	Serpentine minerals, green, blue, or white. snake-skin like appearance	serpentinite
	calcite	marble
	quartz	quartzite

Table 2: Identifying Metamorphic Rocks

ID #	Foliated/Non-Foliated	Crystal Size	Reacts to Acid	Scratches Glass	Rock Name	Other Characteristics

Lab 13
Metamorphic Rocks of Oregon

Name _____ Date _____

Purpose: Explore the metamorphic rocks of Oregon.
Materials: Geologic Map of Oregon
Instructions: Use the provided map to answer the following questions. When asked for a rock unit or formation provide the symbol and name.

1. Where are the majority of metamorphic rocks in Oregon?

2. How old are the majority of metamorphic rocks (what period or era are they from)?

3. What environment do the protoliths (parent rocks) of the metamorphic rocks indicate? Provide some examples.

4. Was the environment mentioned in question 3 part of Oregon (i.e., did the parent rocks form in Oregon)? Why or why not?

5. Are there any metamorphosed rocks in the Coast Range? Suggest a reason why or why not.

6. While they are mountainous, why are there no metamorphic rocks in the Cascade range (or at least no significant units containing them)?

7. What type of metamorphism created the metamorphic rocks of Oregon? What supports this?

Lab 14
Fossils

Name _____ Date _____

Purpose: To examine select fossils and explain how they can be used to study past environments.
Materials: Fossil samples provided by your instructor, Hand lens
Instructions: Using the provided samples, answer the following questions. This is an individual assignment, so while you should work together and discuss your conclusions with each other, you must each turn in your own assignment.

PART I

Pick out 2 fossils from each box provided (4 total) – try to select fossils that are unique from each other in appearance. Please describe the fossils as follows (use a separate sheet of paper if you need to):

1. Given me the fossil number so I know which ones you picked.

2. Tell me the time period represented by your fossil – it is on the box

3. Describe what you see of the original plant/animal. What color is it, how big is, what does it look like. You can also draw a picture if that is easier than describing it.

4. Is it a trace fossil or not – why do you think so?

5. Describe the matrix of the fossil (what is it contained in, if anything). Color, texture, rock type?

6. What type of plant or animal created this fossil?

7. What environment do you think this object was fossilized in?

8. How do you think it was fossilized

9. After you have described all 4 fossils, please place them in the order in which they lived.

PART II

Sample 460 is a fossilized leg bone from a brontotherium, below is a little information on the animal:

1. If you only had this one fossilized piece, speculate on what you might be able to infer about this animal (hint: more than you might think- consider similar living animals)?

For the rest of the questions you may assume that you have found an entire skeleton, think about the description below and speculate on how Paleontologists might have determined this information.

> "The brontotheres (also known as titanotheres) are extinct family of large, rhinoceros-like mammals that were ancestors of the horse, rhinoceros, and tapir. Brontotheres had horn-like structures on their snout; bony knobs protruded from their skull and were covered with thick skin. Males had larger knobs than females. These herbivores ate soft forest vegetation and were up to 8 feet (2.5 m) tall at the shoulder. Brontotheres each had a tiny brain, only as big as a fist. They had four-hoofed toes on each front foot and three-hoofed toes on each rear foot. They lived from the early Eocene until the middle Oligocene (from 58-30 million years ago). Some titanotheres include Brontops (8 ft tall, from North America), Brontotherium (8 ft tall, from North America), Dolichorhinus (4 ft tall, from North America), Eotitanops (1.5 ft tall, from North America and Asia), and Embolotherium (8 ft tall, from Mongolia)." - Jeananda Col, www.enchantedlearning.com

Fossils of these rhinoceros like animals were also found in the Clarno Formation in the John Day region. Pretend you are a Paleontologist – you have studied almost as much biology as geology so you're pretty knowledgeable. Speculate on some of the information presented in the above paragraph.

2. How do scientists know the size of the brain?

3. How do they know what the brontotheres ate?

4. How do they know the time range in which they lived?

5. If you had a complete skeleton of this animal, could you determine something about the environment in which it lived? Explain what you could determine and how.

Lab 15
Discovering Plate Boundaries

Name _____ Date _____

Purpose: To learn about the discovery process, one of the great mysteries of the Earth, using real scientific data.
Materials: Colored pencils
Instructions:

Part 1: Observe data as a specialist.
You will be sorted into four groups and assigned a scientific specialty from the following list:

- Seismologist (specializing in earthquakes)
- Volcanologist (specializing in volcanoes)
- Geochronologist (specializing in seafloor age)
- Geographer (specializing in topography)

All the members of your group should look at the map contained in this lab that is labelled with the specialty you have been assigned. Some maps show locations of events. Other maps show contoured data using colors.

Discuss what your group can observe about the data on your map. For point data (volcanoes and earthquakes), you are looking for distribution patterns. For surface data (topography and seafloor age), you are looking for where the surface is high and where it is low, where it is old and where it is young. Give every group member a chance to discuss what they see. For example, discuss the following questions:

- What sort of data are you looking at?
- What do the different colors indicate?
- What are some examples of extreme data points/locations?
- What are some patterns in the data?
- Where is the data uniform and where is it highly varied?

Now look at the plate boundaries on your group's map. Identify the nature of your data near the plate boundaries. Is it high or low, symmetric or asymmetric, missing or not missing, varying along the boundary or constant along the boundary, etc. As a group, classify the plate boundaries based on your observations of your group's data. Restrict yourselves to about 4–5 boundary types. At this point, do not try to explain the data; just observe!

1. North American Plate _____(classification)

2. Pacific Plate _____

3. African Plate _____

4. South American Plate   _____

5. Eurasian Plate   _____

6. Cocos/Nazca/Caribbean Plates   _____

7. Australian Plate   _____

8. Antarctic Plate   _____

9. Indian Plate   _____

10. Arabian Plate   _____

Now, assign a colored pencil color to each boundary type in your classification scheme. Color your first Plate Boundary Map to locate your group's boundary types. If the data are asymmetric at a particular boundary type, devise a way of indicating that on your plate boundary map. Each person should mark the boundary types identified by the group on their own map. Each person should write down descriptions of the group's plate boundary classifications on the back of their map.

Part 2: Specialists assemble!

Each member of your group is now a specialist in your particular field of studying plate boundaries. For the next part, you'll be split apart and act as one specialist among other specialists to work together to study a specific plate.

To get started, each specialist group will be divided into equal groups of different specialists, one specialist of each type per group. Each person should make a brief presentation to the rest of their group about their Scientific Specialty's data and classification scheme. This new group will be assigned to a specific plate or plates, working together to provide information about their plate from each specialty.

Next, compare the classifications of boundary type for your plate based on each type of data. Are there common extents (along the boundaries) between the different classifications? Can your plate group come up with a new classification scheme that now includes data from all four Scientific Specialties? As above, assign a color to represent each of your plate boundary types. If a boundary is asymmetric, be sure to devise a way to represent the asymmetry. Mark the boundaries of your plate or plate grouping using your color scheme on your second Plate Boundary Map. Also, write a description of the plate boundary classes you have used in the space below.

Field Trip 1
Downtown Salem Walking Tour

Name _____   Date _____

The purpose of this trip is to see a variety of rocks used as commercial building stone in Salem. Some are nearly identical to rocks you've seen in lab; others will be new. Building stones are selected mainly for their beauty, and are not normally integral to the building structures. As you'll learn, many of these rocks were quarried from faraway places. Read the descriptions and answer the questions at each stop. Remember to use your hand lens!

Begin at Allann Bros. Coffee Bistro (The Beanery) at 220 Liberty Street NE, Salem

STOP 1: 248 LIBERTY (CAFÉ SHINE)

Find a surface where the red-paint on the bricks here has been removed.

- Is this rock igneous, sedimentary, or metamorphic? _____

- What is the name of this rock? _____

STOP 2: 260 LIBERTY STREET

Identify the main mineral present in the rock (the largest crystals).

STOP 3: 255 LIBERTY STREET

- Is this rock igneous, sedimentary, or metamorphic? _____

- What is the texture of this rock? _____

- What is the composition of this rock? _____

- What is the rock name? _____

STOP 4: 241 LIBERTY STREET

This rock is fake. But what is it imitating?

- Is this "rock" igneous, sedimentary, or metamorphic? _____

- What is the texture of this "rock"? _____

- What is the "rock" name? _____

STOP 5: 225 LIBERTY STREET (JACKSON'S JEWELER)

- Is this rock igneous, sedimentary, or metamorphic? _____

- What is the texture of this rock? _____

- What is the composition of this rock? _____

- What is the rock name? _____

- This rock has several dikes running through it. Choose one and compare its texture and composition to the main rock.

STOP 6: STARBUCKS

- This rock is foliated. What does this mean?

- What is the red mineral in this rock? _____
- What is this rock's name? _____

Turn west (right) onto Court Street.

STOP 7: 377 COURT STREET (INDIA PALACE RESTAURANT FLOWER PLANTER)

- Are the stones on this planter igneous, sedimentary, or metamorphic? _____

- What is the rock name? _____

- What previous stop do these rocks resemble?

Continue down Court Street, cross and turn left Commercial Street.

STOP 8: 179 COMMERCIAL STREET

Look down at the sidewalk. Notice the blocks of purple glass in the pavement.

- What fracture pattern do these blocks exhibit?

STOP 9: 129 COMMERCIAL STREET

This is an old rock, at least half a billion years old.

- Is it igneous, sedimentary, or metamorphic? _____
- What is the rock name? _____
- What is the mafic mineral in this rock? _____
- What is a possible parent rock? _____

STOP 10: PIONEER TRUST BANK

This rock is a good example the schiller effect. The flash of color you see across the feldspar crystals.

- Which feldspar is present in this rock? _____
- Notice that the feldspar are zoned (the sun's reflection on the feldspars will help see this). What element – sodium, calcium, or potassium, would most likely be present in the rims of these feldspars? Explain why.

Turn left State Street, away from the river (back towards Liberty Street).

STOP 11: 379-383 STATE STREET (MA VALISE JEWELRY STORE)

- Is this rock igneous, sedimentary, or metamorphic? _____
- This rock is soft, but does not react to acid. What is the rock name? _____
- This rock is ultramafic, what is the parent rock? _____

Cross Liberty and State Street.

STOP 12: 416 STATE STREET (KEY BANK)

The bricks in this building are composed of two types of limestone: coral limestone and travertine (non-marine limestone).

- How are the two limestone types similar?

- What evidence can you find that indicates these bricks are composed of calcite?

Continue down State Street and cross the street at High Street.

STOP 13: CORNER OF STATE AND HIGH STREETS
(512-516 STATE STREET – THE MARION COUNTY COURTHOUSE)

This rock is made out of the same mineral as the previous stop, but is not limestone.

- What is the rock name? _____

- Look closely with a hand lens. What light-colored mineral forms thin streaks in the rock?

- What is the metallic mineral here?

Cross and continue down State Street (away from Liberty). Follow the sidewalk into Willamette University towards the large stand of Redwoods.

STOP 14: CORNER OF STATE AND COTTAGE STREETS
(700 STATE ST.) – HALLIE FORD ART MUSEUM

Find the rock that lines the base of the building.

- Sketch the rock below.

♦ Describe how the texture of this rock is different from the gabbro that you have previously seen today?

Cross and continue down State Street (away from Liberty). Follow the sidewalk into Willamette University towards the large stand of Redwoods.

STOP 15: WILLAMETTE UNIVERSITY (ERRATIC FROM MISSOULA FLOODS WITH PLAQUE)

This boulder was deposited in Oregon 13,000-11,500 years ago (The plaque is lying to you) as part of the Missoula. It is called a flood erratic. Similar to Erratic Rock outside McMinnville.

♦ Identify the mafic mineral in this rock. _____

STOP 16: WILLAMETTE UNIVERSITY, EATON HALL

The foundation of this building is a tuffaceous sandstone

♦ How is the weathering of this rock different than the sandstone at stop 1?

♦ Why would the weathering of this rock be different if they are both sandstone?

Cross State street toward the Capitol Building, go around to the grassy area (away from Liberty).

STOP 17: STATE CAPITAL BUILDING, CIRCUIT RIDER STATUE

♦ Is this rock igneous, sedimentary, or metamorphic? _____

- Does it contain quartz crystals? (You may see this better if the rock is wet.) _____

- Look closely with a hand lens to identify the mafic mineral: _____

- What is the rock name? _____

Continue around towards the front of the Capital Building.

STOP 18: STATE CAPITAL BUILDING, JASON LEE AND/OR JOHN MCLAUGHLIN STATUES

- What rock texture do you see here that was not at the Circuit Rider statue? _____

STOP 19: STATE CAPITAL BUILDING, VENT STACK BEHIND THE MCLAUGHLIN STATUE

- Is this rock metamorphic or sedimentary? _____

- What is the texture of this rock?

- What is the name of this rock? _____

Continue around towards the front of the Capital Building.

STOP 20: STATE CAPITA BUILDING, OREGON VETERANS MEDAL OF HONOR MEMORIAL

- The rock on this memorial is very similar to the rock of the previous statues. What about them indicates that this memorial is much more recent (other than the date on the plaque)?

Continue along the front of the Capital Building.

STOP 21: STATE CAPITAL BUILDING, OREGON TRIBES PLAQUES

- What is the rock name? _____

- What mineral is present in these rocks that has not been in the previous 3 stops?

Continue west across the park to the cast iron fountain at the far end, next to Cottage Street.

STOP 22: STATE CAPITAL BUILDING, SMALL ROUND FOUNTAIN NEXT TO COTTAGE STREET

The pavers around the fountain are made of limestone. Some of them are travertine as we saw at Key Bank.

- What feature distinguishes the travertine bricks from the other limestone?

- Notice some of the limestone bricks have a zig-zag pattern running through the rock. Suggest an origin for this feature.

Field Trip 2
Oregon Coast

Name _____ Date _____

The purpose of this trip is to explore and identify some of the rocks found along the Oregon Coast. Be sure to wear sturdy shoes (no sandals). Remember: you are representing Chemeketa Community College, all college rules and policies apply even while on this trip.

For Sketches: When drawing your sketch or profile, be sure to make it to scale. Always include that scale in your picture. Label all rock units as well as any prominent features as indicated by your instructor. The goal is not artistic ability, but accurate representation of the outcrop and any processes occurring at the location.

STOP 1: EDDYVILLE TURBIDITE

- Sketch the outcrop.

- Where were these rocks actually deposited, prior to being uncovered in a road cut?

- Describe how this sequence formed.

STOP 2: SEAL ROCK STATE RECREATION SITE

Look at the various outcropping of rock.

- What is this site composed of (there should be more than one answer here)?

- How was the rock here deposited and in what order?

Elephant Rock has been radiometrically dated to be approximately 14 million years old and is related to the volcanic activity that produced the Columbia River Flood Basalts.

- Sketch Elephant rock.

- What are the distinctive linear, verticals cracks in the basalt?

STOP 3: NYE BEACH

- Sketch the outcrop of Jump-off Joe.

- Why this outcrop is so prone to landslides?

STOP 4A: YAQUINA LIGHT HOUSE NATURAL AREA

Look at the quarry walls.

- What type of rock is present here? _____
- Sketch a small section showing the splayed columnar jointing.

- Why do these shapes form rather than the simple vertical columns like we saw at Elephant Rock?

- What do you think this rock may have been used for?

STOP 4B: COBBLE BEACH

- What type of rock makes up Cobble Beach? _____
- Identify any phenocrysts.

- How would you rate the sorting and rounding of these cobbles? _____
- What type of rock would form if these cobbles were cemented together? _____

STOP 5: BEVERLY BEACH

- Find at least two fossils and sketch them.

STOP 6: CAPE FOULWEATHER

- What type of rock are we standing on? _____

- How thick would you estimate this flow to be? _____

- Where do you think this deposit came from and why was it able to travel so far?

STOP 7: DEPOE BAY

- There are two types of rocks here. What are they?

- Sketch the structures seen in the outcrop here.

- How do these structures form?

- There is a rind on these formations, why did it form?

Field Trip 3
Shellburg and Henline Falls

Name _____ Date _____

The purpose of this trip is to explore and identify some of the rocks and geologic processes occurring in the Oregon Western Cascades. Be sure to wear sturdy shoes (no sandals). Remember: you are representing Chemeketa Community College, all college rules and policies apply even while on this trip.

For Sketches: When drawing your sketch or profile, be sure to make it to scale. Always include that scale in your picture. Label all rock units as well as any prominent features as indicated by your instructor. The goal is not artistic ability, but accurate representation of the outcrop and any processes occurring at the location.

STOP 1: SHELLBURG FALLS

- Our first stop not even rock yet! However, we are standing in a geologic feature that can create a rock all its own. Describe, in detail, the topography of what we are standing upon. What type of process can be seen here and what is a possible rock that this geologic process could create?

- What type of rock do we see here? Describe the rock in terms of composition, morphology (what does it look like *in situ*), texture, etc. How do you think it got here and where did it come from?

- At the upper falls you can see there is some interesting erosion going on. Why do you think there is a very clear overhang of rock that the falls is flowing over? What does that suggest about the strength of the rock beneath the overhang?

- Get up close to the contact between the interbed and the basalt flow we seen here. Describe the appearance of the rock. What type of metamorphism occurred here?

STOP 2: HENLINE FALLS

- What type of rock do we see here? Describe the rock in terms of composition, morphology (what does it look like *in situ*), texture, etc. How do you think it got here and where did it come from?

- Choose 3 different types of rocks from the debris flow here. Describe them below using mineralogy, color, texture. Come up with a possible scenario about how and where it initially formed.

    1.

    2.

    3.

STOP 3: THREE POOLS

- Check out the topography here. Why do you think we see the features here? Also, note the beach. We are going to do a little panning here. Why do you think this might be a good place to pan?

Printed in the USA
CPSIA information can be obtained
at www.ICGtesting.com
JSHW061001060824
67508JS00003B/16